Patterns Al

Written by Margie Burton, Cathy French, and Tammy Jones

Look!

Here are patterns

on the butterfly.

Look!

Here are patterns

in the garden.

Look!

Here are patterns

on the spider web.

Look!

Here are patterns
on the giraffe.

Look!

Here are patterns

on the tiger.

Look!

Here are patterns
on the zebra.

Look!

Here are patterns
on the ball.

Here are patterns

on me.

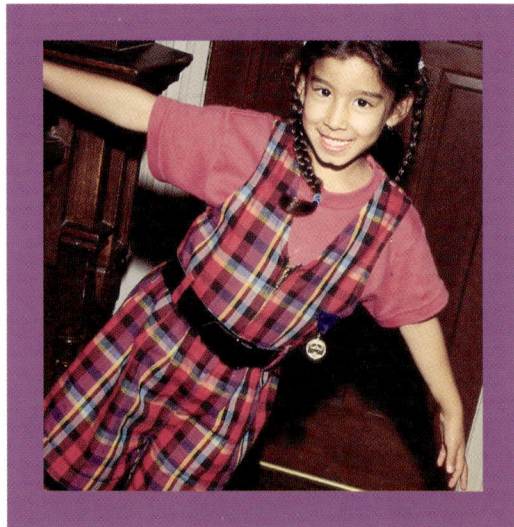